THE LITTLE ENGINEER COLORING BOOK

HOW TO BUILD A HOUSE

SETH MCKAY

The Little Engineer Coloring Book: How to Build a House by Seth McKay
www.TheLittleEngineerBooks.com

Copyright © 2018 by Seth McKay

All rights reserved. No portion of this book may be reproduced, stored in a retrieval system, or transmitted in any form or by any means electronic, mechanical, photocopy, recording, scanning, or other-except for brief quotations in critical reviews or articles, without the prior written permission of the publisher.

Creative Ideas Publishing titles may be purchased in bulk for educational, business, fund-raising, or sales promotional use. For information, please email permissions@TheLittleEngineerBooks.com.

ISBN-13: 978-1-952016-02-8

Published by: Creative Ideas Publishing

Table of Contents

Introduction (*for Parents*) v

Introduction (*for the Little Engineers*) vi

Tips on Using this Book vii

Survey the Land 1

Prepare the Land 2

An Architect Draws the House 3

Concrete Trucks Help Make the Foundation 4

A Truck Delivers Building Supplies 5

Walls are Built 6

How to Hold Walls Together 7

Add Windows 8

Add the Roof 9

Add Water Pipes 10

Add Wires for Electricity 11

Cover the Walls, Pipes and Wires 12

Add an Air Conditioner 13

Now the House is Ready! 14

Where Does the Electricity Come From? 15

Fuel Power Plant 16

Moving Water Power Plant 17

Sun Power Plant 18

Wind Power Plant 19

Making Power at Your House 20

Where Does Water Come From? 21

How the Water is Cleaned 22

Well Water 23

House Complete! 24

Hot Water Heater 25

Refrigerator 26

Oven 27

Dishwasher 28

Blender 29

Washing Machine 30

Dryer 31

Outdoor Grill 32

Certificate of Completion 33

BONUS: Special Preview of Cars and Trucks Coloring Book 34

Introduction *(for parents)*

Thank you so much for your interest in this book. In these books, your child will be introduced to new, interesting topics and begin to understand how things work.

I've always wanted my children to understand how things work, and in general, understand that objects are not magical boxes that work for some unknown reason. I want them to understand that objects are quite simple when broken down into a few key components. Without any direction, it is possible for a child to get the wrong understanding of objects that we consider to be simple.

For example, we understand that a printer has a roller that moves the paper and a little ink jet that sprays the paper with ink as it rolls by to make images. However, it would be very easy for our children to see the same printer and conclude that this magic box shoots out paper, and this is where paper comes from.

I chose to make this book a coloring book because I wanted to ensure that the book was also fun! I don't want this book to be too serious, boring or overwhelming. Coloring helps keep the book fun while incorporating new and challenging topics.

Book Objective:

The objective of this book is not to make your child an engineer (I hope your child chooses whatever career path suits him or her best), and the objective is not for your child to know exactly how everything works and memorize each step of building a house.

The objective is simply to begin to introduce your child to new, interesting concepts that might catch his or her attention and help your child understand that seemingly complex things are just a few simple items put together.

Introduction *(for the Little Engineers)*

Hello Little Engineer!

My name is Seth, and I'm the Chief Engineer of this book. I'm so happy you have this book so we can learn how to build a house together. Of course, we are going to have fun too!

We will start with an empty field, then build a nice big house, discover where the water comes from that ends up in our bath tub, and even learn how appliances work.

Enjoy coloring, ask your parents lots of questions, and most of all, have fun!

Tips on Using this Book

- **Read the book as a story to your child** – Before coloring the book, read it one time as a full story to help your child get a sense of the "big picture" before they focus on a single page to color.

- **Add your own explanations** – You know your child best so take the wording on a page and expand on it to help your child understand.

- **Relate the pages to your surroundings** – Not all houses are the same, so a picture might not match your house so be sure to point out how this house is different from your house. For example, you may have a metal roof, but this book mentions shingles.

- **Take a field trip** – Stop by a half built neighborhood or point out objects like water towers. This will allow your child to understand that these are real objects that we use everyday.

LET'S GET STARTED!

SURVEY THE LAND

THE LAND SURVEYOR HELPS FIND THE PERFECT SPOT FOR A NEW HOUSE TO BE BUILT.

THE SURVEY LEVEL TOOL HELPS THE LAND SURVEYOR MEASURE HOW TALL A HILL IS.

The Little Engineer Coloring Book: How to Build a House

PREPARE THE LAND

BULLDOZER

A BIG BULLDOZER
SMOOTHES AND FLATTENS THE DIRT.

AN ARCHITECT DRAWS THE HOUSE

AN ARCHITECT DESIGNS THE HOUSE AND MAKES BLUEPRINTS TO GIVE TO THE BUILDERS.

BLUEPRINTS ARE THE INSTRUCTIONS THAT SHOW THE BUILDERS HOW TO BUILD THE HOUSE.

CONCRETE TRUCKS HELP MAKE THE FOUNDATION

BUILDERS USE THE BLUEPRINTS AND MAKE THE FOUNDATION OF THE HOUSE TO MATCH THE BLUEPRINT DRAWINGS. THE FOUNDATION IS MADE OF CONCRETE. A BIG CONCRETE TRUCK COMES TO THE HOUSE AND POURS CONCRETE TO MAKE THE FOUNDATION.

A TRUCK DELIVERS BUILDING SUPPLIES

A TRUCK DELIVERS BUILDING SUPPLIES TO THE HOUSE.
BIG PIECES OF WOOD ARE DELIVERED
AND MOVED OFF THE TRUCK
USING A CRANE OR FORKLIFT.

WALLS ARE BUILT

THE BUILDERS USE PIECES OF WOOD TO BUILD WALLS ON THE FOUNDATION.
THE TALL PIECES OF WOOD IN THE WALL ARE CALLED STUDS.
THERE IS A TOOL CALLED A "STUD FINDER" THAT CAN HELP YOU FIND STUDS INSIDE YOUR OWN WALLS.

HOW TO HOLD WALLS TOGETHER

THE BUILDERS USE SCREWS OR NAILS TO HOLD THE WALLS TOGETHER.

ADD WINDOWS

BUILDERS LEAVE BIG SQUARE OR RECTANGULAR HOLES IN THE WALLS;
LATER THEY INSTALL WINDOWS IN THESE HOLES.
WINDOWS ARE MADE AT A FACTORY, AND DELIVERED TO THE HOUSE READY TO FIT INTO THE WALLS.

ADD THE ROOF

A CRANE LOWERS THE ROOF ONTO THE HOUSE.
A WATERPROOF SHEET IS THEN PUT
ON TOP OF THE ROOF,
AND SHINGLES ADDED ON TOP OF THAT.

The Little Engineer Coloring Book: How to Build a House

ADD WATER PIPES

A PLUMBER ADDS PIPES INSIDE THE WALLS FOR WATER. THE PIPES MOVE WATER TO DIFFERENT ROOMS INSIDE THE HOUSE. BATHROOMS, KITCHENS AND LAUNDRY ROOMS ALL NEED WATER PIPES.

ADD WIRES FOR ELECTRICITY

AN ELECTRICIAN ADDS WIRES IN THE WALLS. THESE WIRES CARRY ELECTRICITY TO THE LIGHTS, OUTLETS, FANS, REFRIGERATOR AND ANYTHING ELSE THAT NEEDS ELECTRICITY.

COVER THE WALLS, PIPES, AND WIRES

THE INSIDE OF THE WALLS ARE COVERED BY
BIG FLAT SHEETS CALLED SHEETROCK.
SCREWS HOLD THE SHEETROCK UP ONTO THE WALLS;
THE SCREWS ARE THEN HIDDEN BY COVERING THEM
WITH 'SHEETROCK MUD' AND PAINT.

ADD AN AIR CONDITIONER

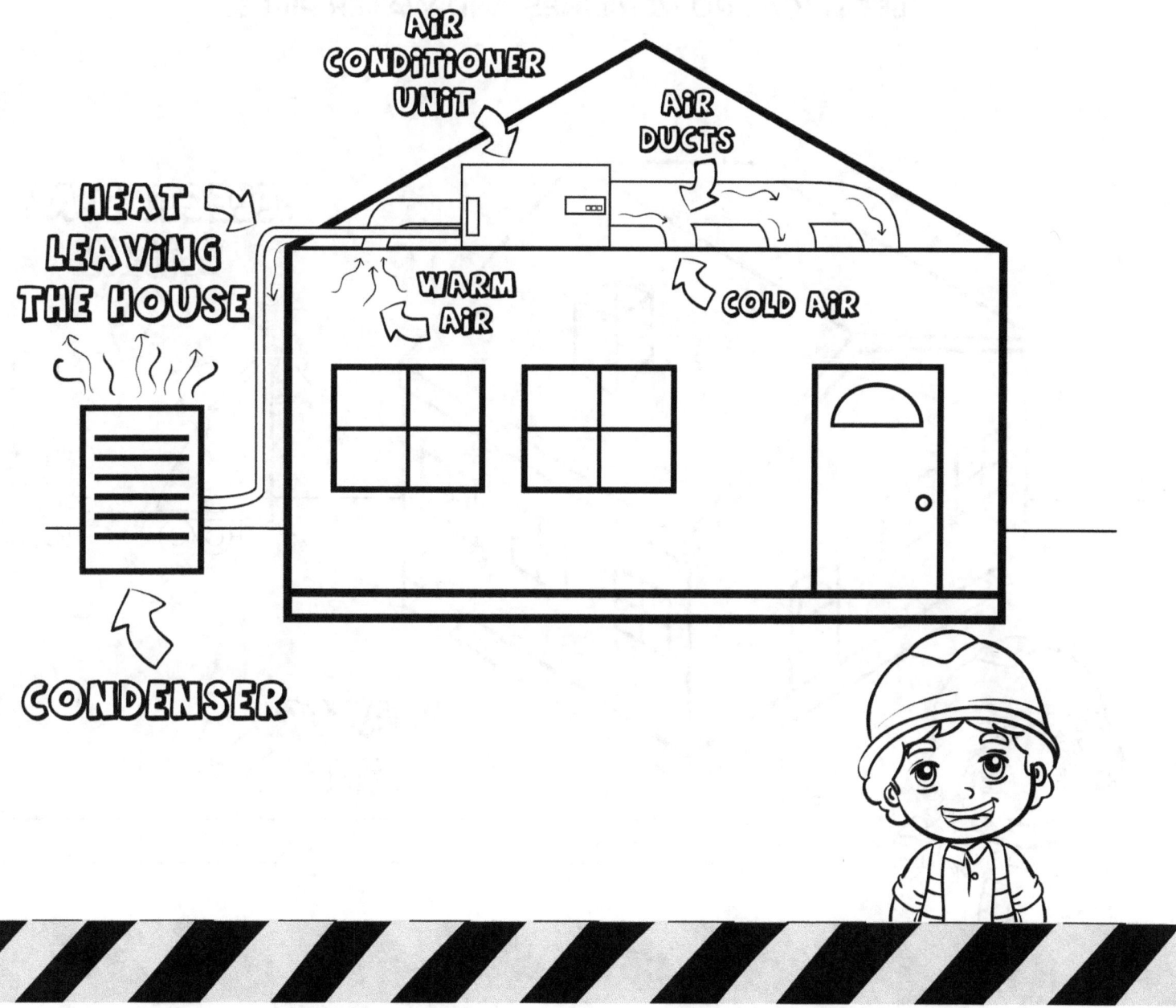

AIR CONDITIONERS HELP KEEP THE AIR INSIDE THE HOUSE FROM GETTING TOO HOT OR COLD. ON A HOT DAY, THE AIR CONDITIONER PULLS HEAT OUT OF THE AIR AND SENDS IT OUT OF THE HOUSE; THIS MAKES THE AIR INSIDE THE HOUSE NICE AND COOL.

The Little Engineer Coloring Book: How to Build a House

NOW THE HOUSE IS READY!

LET'S CONNECT THE HOUSE TO THE TOWN'S ELECTRICAL POWER LINES AND WATER PIPES.

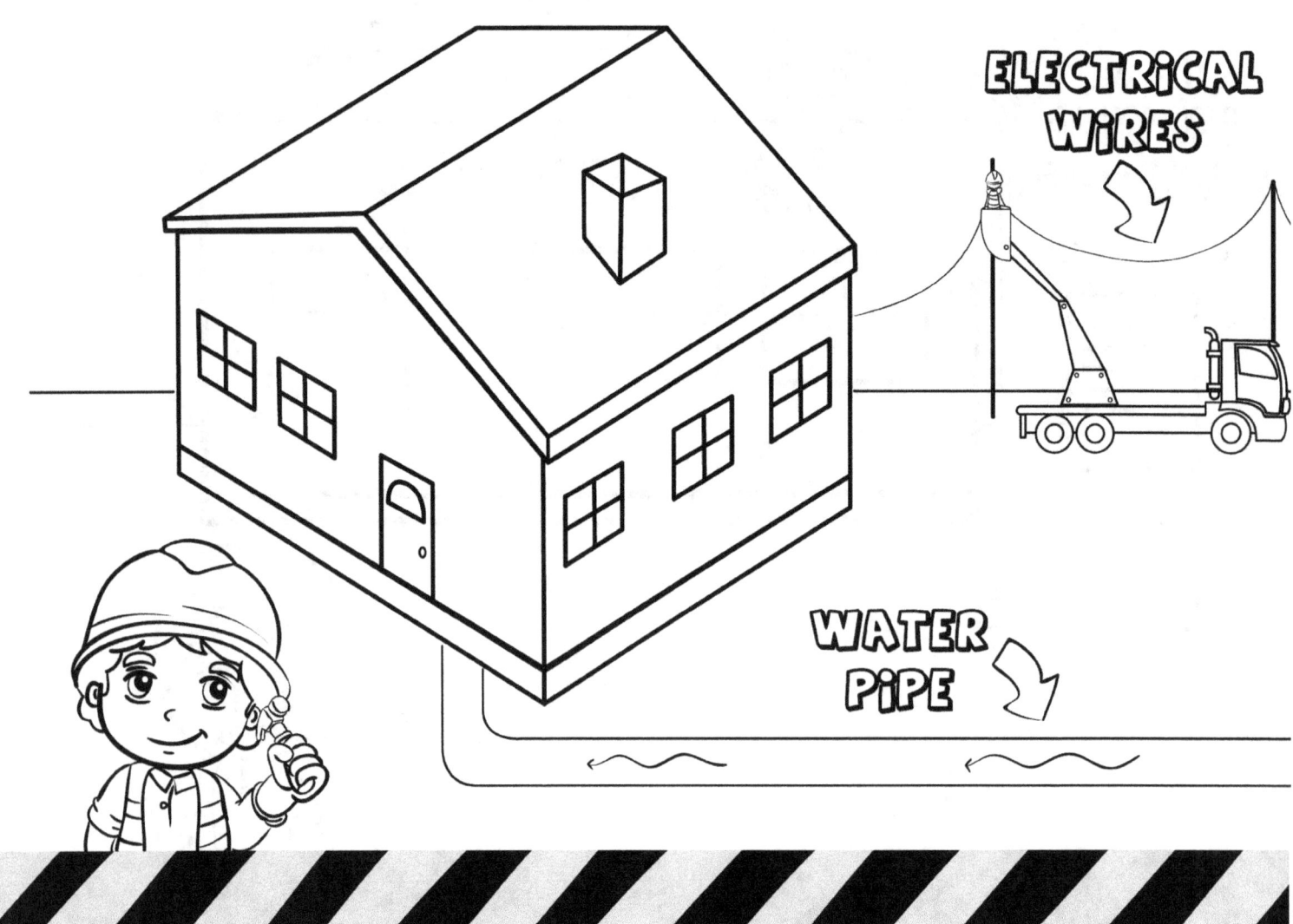

THE POWER COMPANY CONNECTS THE HOUSE'S ELECTRICAL WIRES TO TALL POWER LINES. SOMETIMES THESE POWER LINES ARE UNDERGROUND. THE WATER COMPANY CONNECTS THE HOUSE'S WATER PIPES TO A BIG WATER PIPE THAT IS BY THE ROAD.

WHERE DOES THE ELECTRICITY COME FROM?

HUGE FACTORIES CALLED "POWER PLANTS"
MAKE ELECTRICITY FOR THOUSANDS OF HOUSES!
THE NEXT FOUR PAGES WILL SHOW
HOW THE POWER PLANTS MAKE ELECTRICITY.

FUEL POWER PLANT
GAS TURBINE AND BOILER POWER PLANT

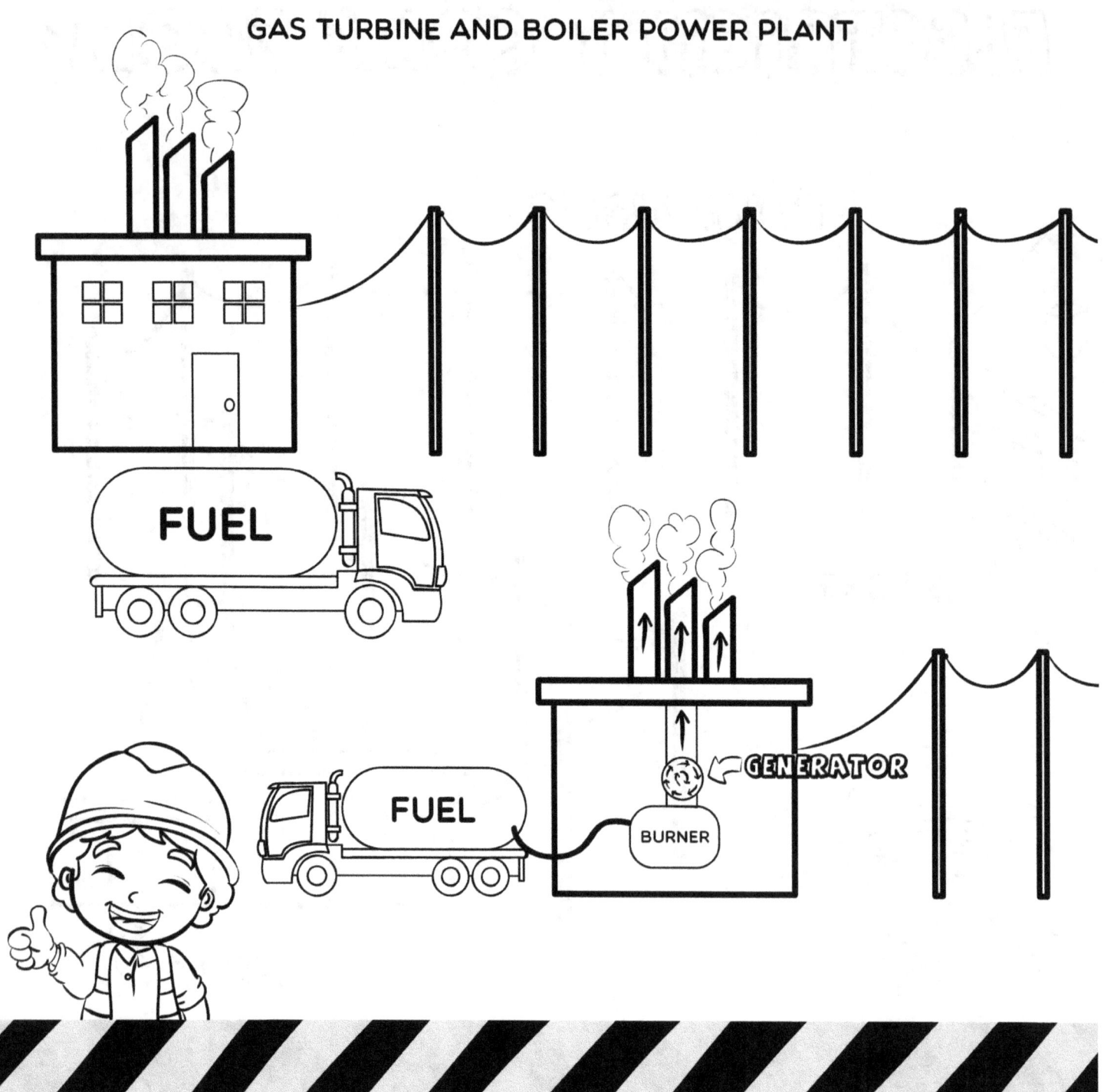

THE POWER PLANT BURNS FUEL TO SPIN A GENERATOR. THE SPINNING GENERATOR MAKES ELECTRICITY.
SOMETIMES WATER IS HEATED UP TO MAKE STEAM, AND THAT STEAM IS STRONG ENOUGH TO SPIN THE GENERATOR.

MOVING WATER POWER PLANT

HYDROELECTRIC POWER PLANT

A DAM IS A BIG WALL THAT STOPS WATER AND MAKES A LAKE. THE BIG WALL HAS A PIPE INSIDE TO LET THE WATER FALL DOWN THE PIPE AND THE FALLING WATER SPINS A GENERATOR TO MAKE ELECTRICITY.

SUN POWER PLANT
SOLAR PANEL POWER PLANT

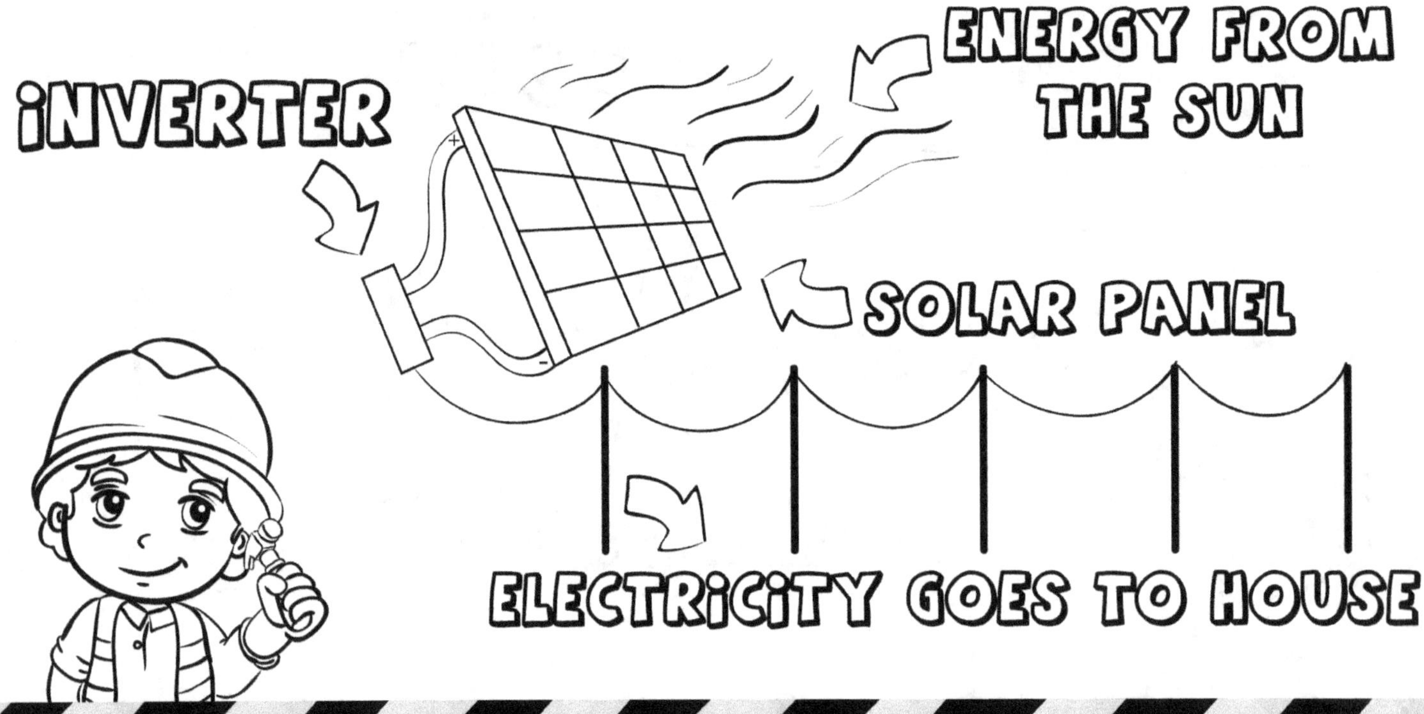

SUN POWER PLANTS CAN HAVE THOUSANDS
OF SOLAR PANELS.
SOLAR PANELS TAKE THE BRIGHT ENERGY
FROM THE SUN AND CHANGE IT INTO ELECTRICITY.

WIND POWER PLANT
WIND TURBINE POWER PLANT

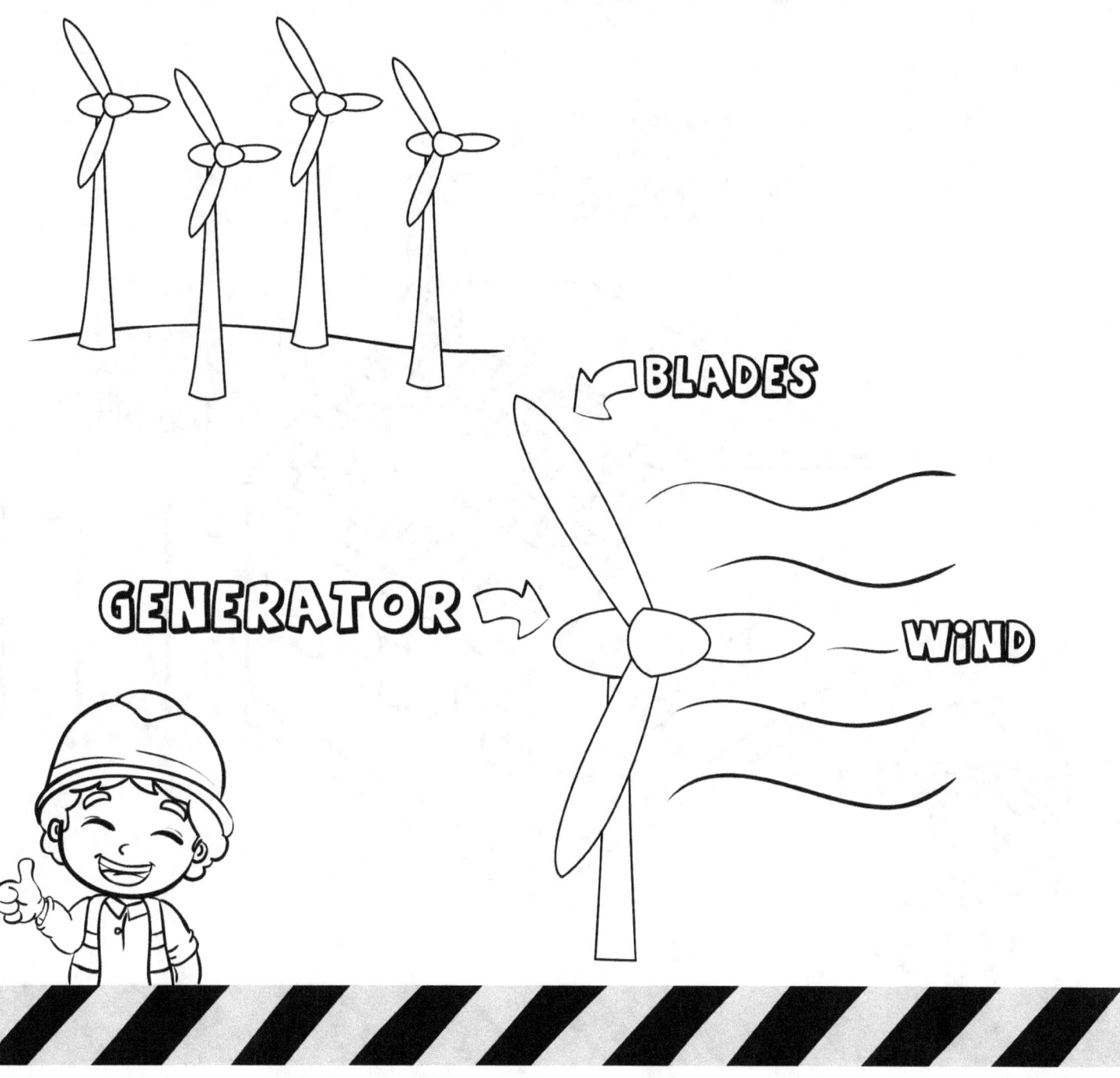

A WIND TURBINE CAN BE TALLER
THAN A 30 STORY BUILDING!
WIND PUSHES THE BLADES AROUND WHICH SPINS
THE GENERATOR TO MAKE ELECTRICITY.

MAKING POWER AT YOUR HOUSE

ELECTRICITY CAN BE MADE AT YOUR HOUSE TOO!
SOLAR PANELS AND WIND TURBINES
CAN BE INSTALLED AT YOUR HOUSE;
IT IS LIKE HAVING YOUR OWN TINY POWER PLANT!

WHERE DOES WATER COME FROM?

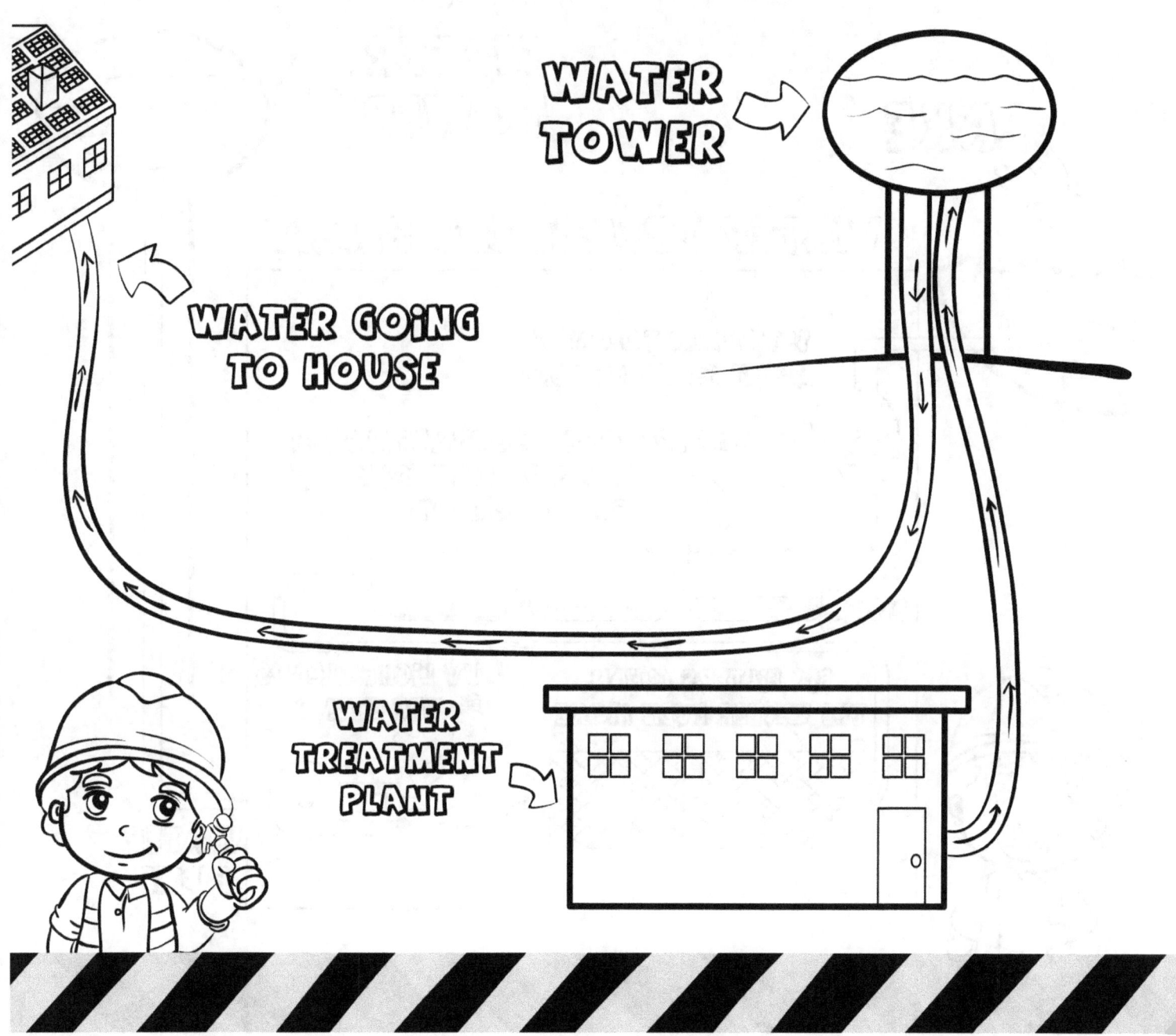

THE HOUSE'S WATER PIPES ARE CONNECTED TO TALL WATER TOWERS. THE WATER FALLS DOWN THE TOWER AND INTO THE HOUSE WHEN THE WATER IS TURNED ON; THIS IS WHY WATER COMES OUT OF THE FAUCET OR SHOWER QUICKLY.

HOW THE WATER IS CLEANED

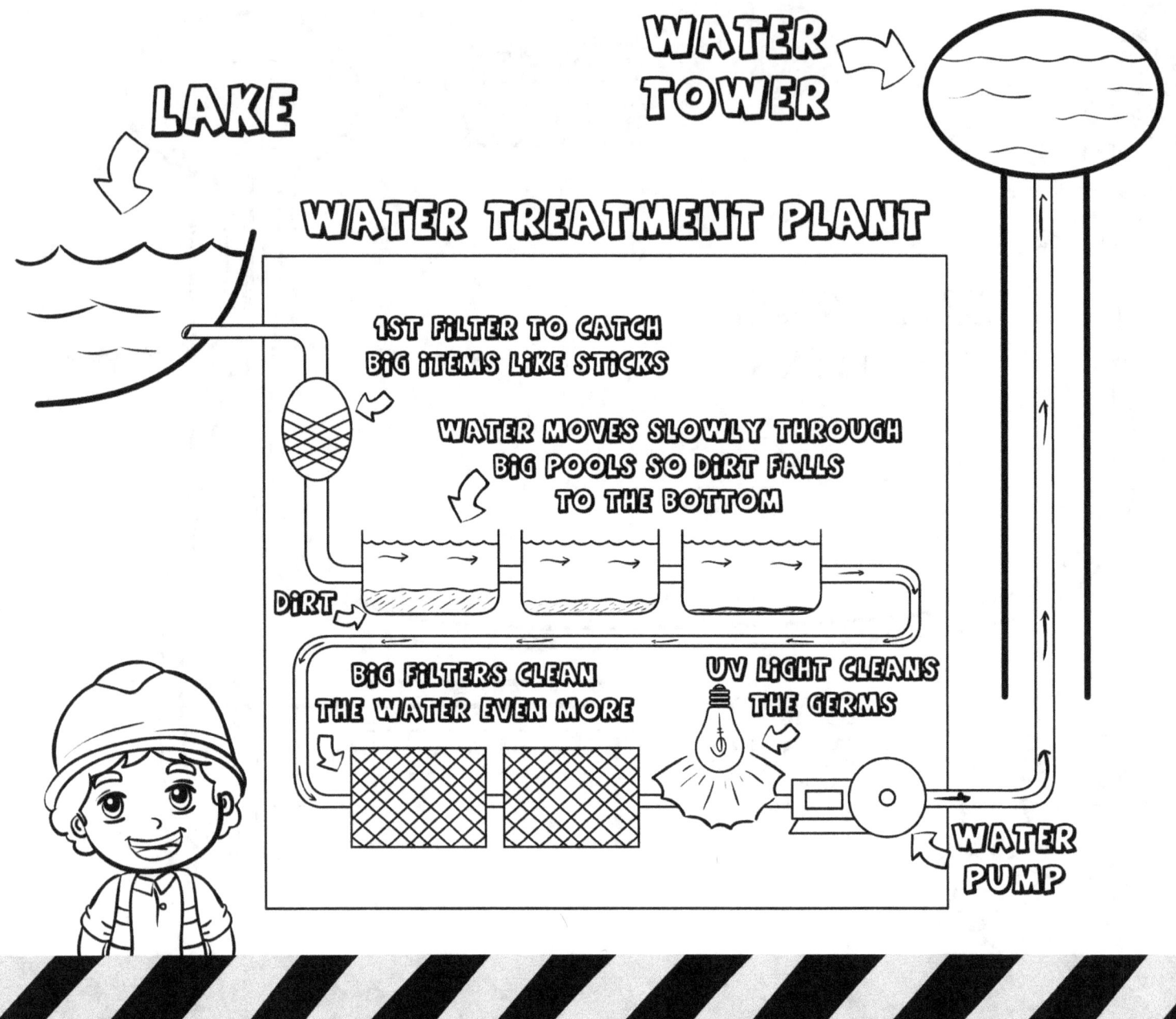

THE WATER TREATMENT PLANT CLEANS ALL THE WATER
TO MAKE IT SAFE TO DRINK.
A WATER PUMP THEN PUMPS ALL THE WATER
UP INOT THE WATER TOWER.

WELL WATER

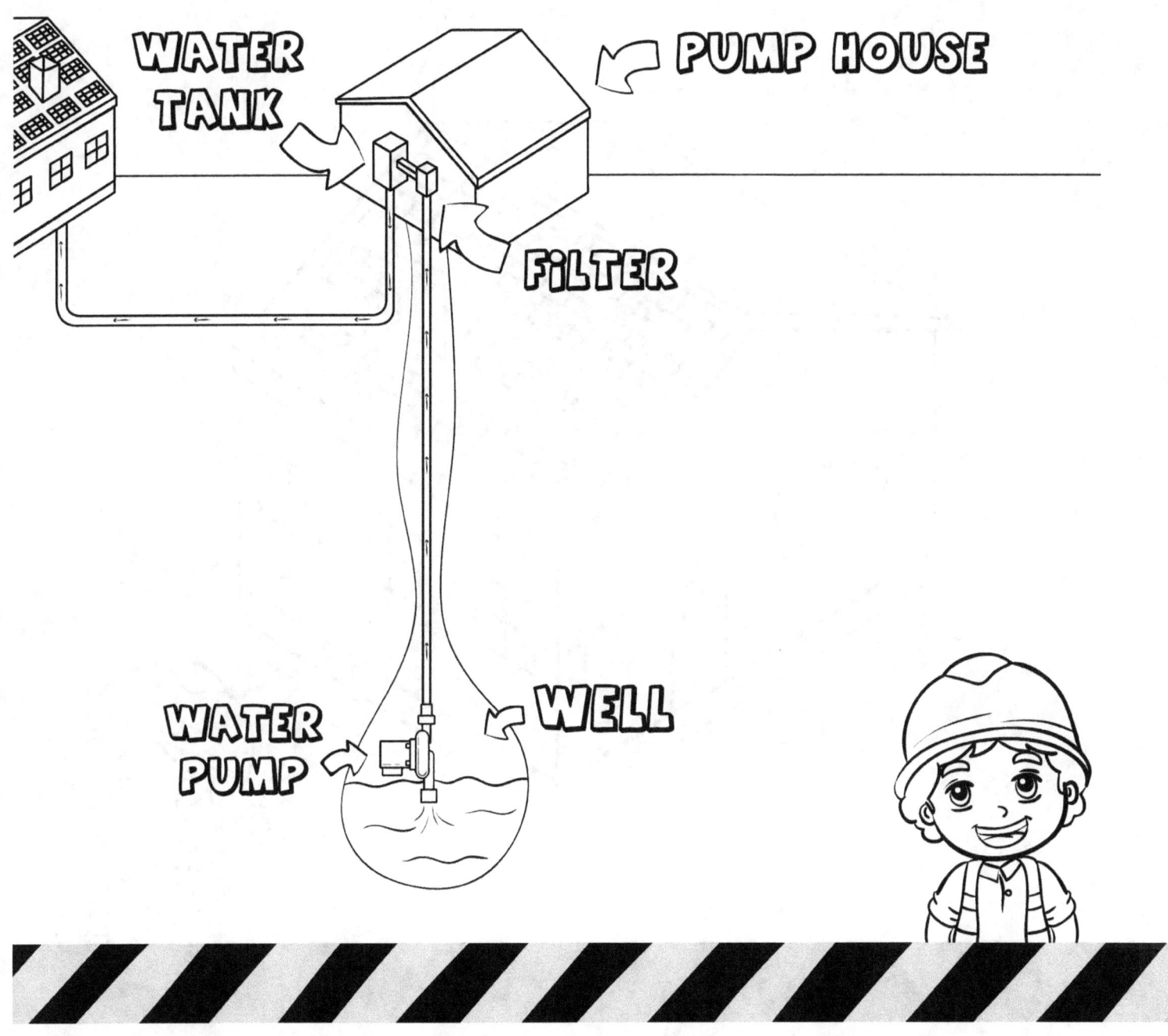

SOME HOUSES ARE NOT CONNECTED TO A WATER TOWER SO THESE HOUSES HAVE A DEEP HOLE CALLED A WELL. A PUMP PUSHES WATER OUT OF THE WELL, THROUGH A FILTER, AND THEN THE WATER IS PUT INTO A WATER TANK.

HOUSE COMPLETE!

THE HOUSE IS COMPLETE, AND WE KNOW HOW THE WATER AND ELECTRICITY GET TO THE HOUSE.
NOW LET'S ADD A FEW MACHINES TO MAKE LIFE EASIER.
THESE MACHINES ARE CALLED APPLIANCES.

HOT WATER HEATER

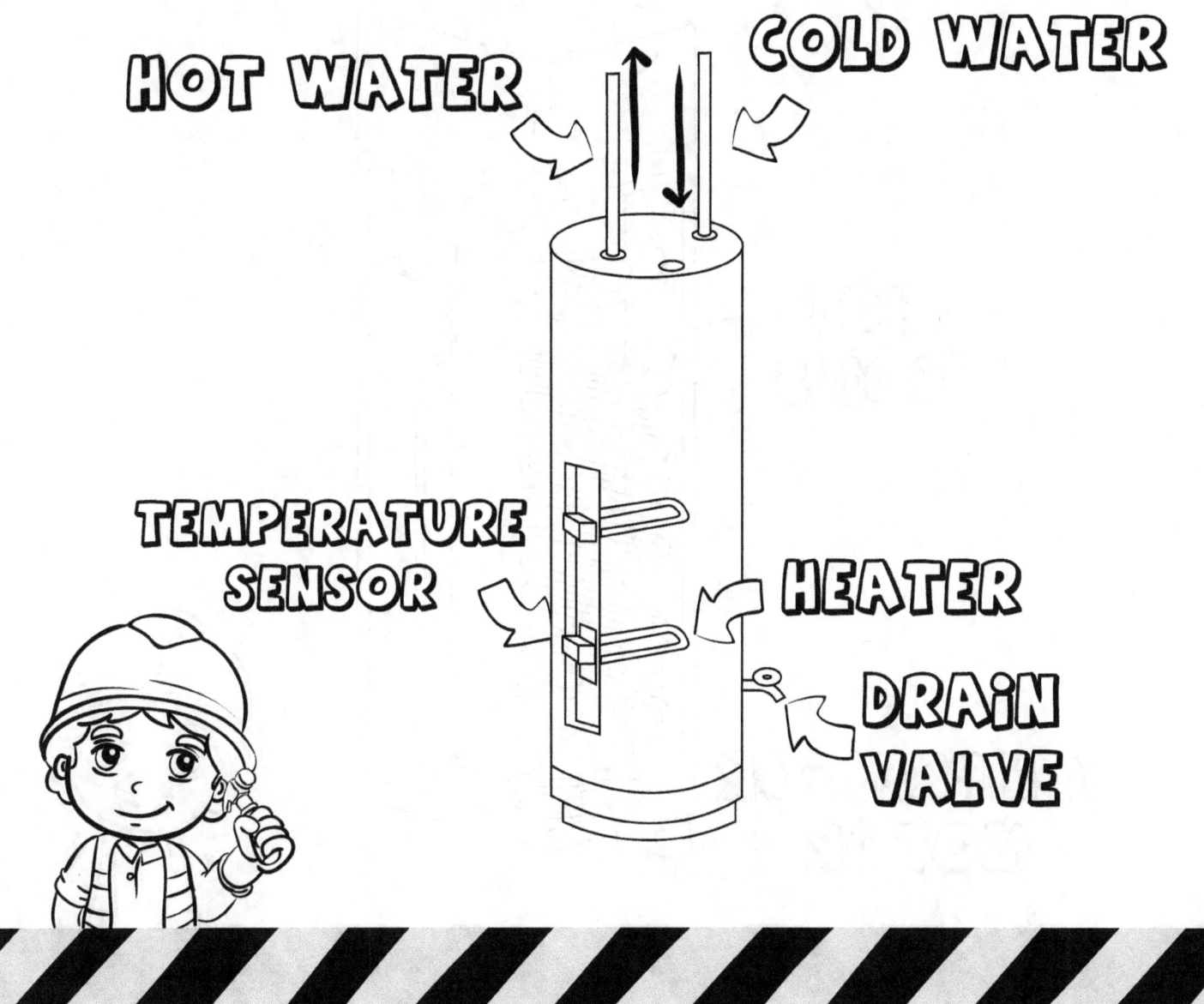

THE HOT WATER HEATER MAKES THE WATER HOT SO THAT A BATH OR SHOWER CAN HAVE HOT WATER. WITHOUT IT, ALL THE WATER IN YOUR HOUSE WOULD BE COLD LIKE THE WATER THAT COMES OUT OF THE HOSE OUTSIDE. THE HOT WATER HEATER IS A BIG TANK OF WATER AND A HEATER MAKES ALL THE WATER IN THE TANK HOT. THERE ARE ALSO HOT WATER HEATERS THAT DON'T HAVE A TANK OF WATER.

REFRIGERATOR

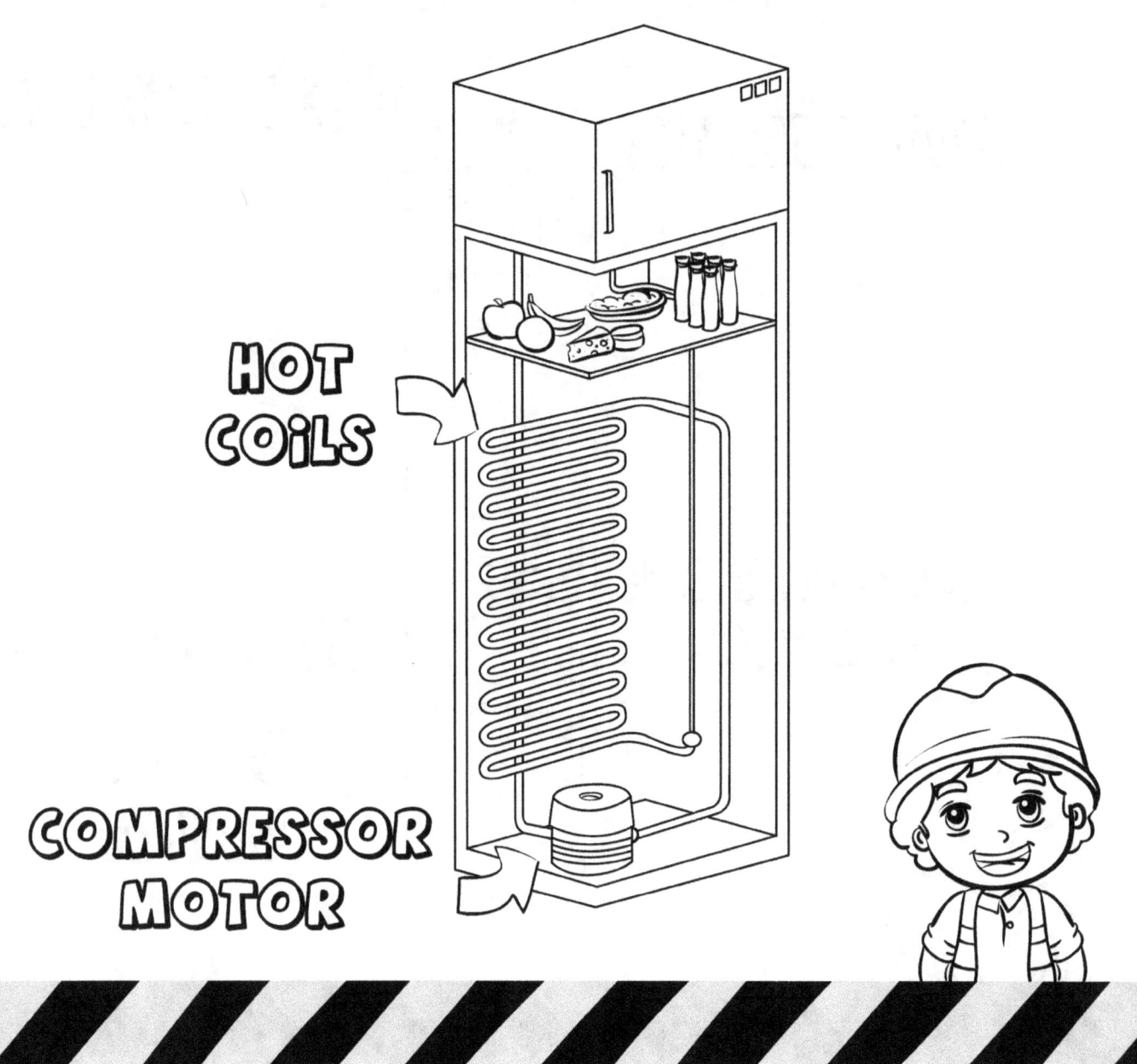

THE REFRIGERATOR KEEPS FOOD AND DRINKS COLD.
A REFRIGERATOR WORKS JUST LIKE
THE AIR CONDITIONING SYSTEM FOR THE HOUSE.
THE REFRIGERATOR PUMPS A SPECIAL FLUID,
CALLED REFRIGERANT, THROUGH THE PIPES.
THE PIPES MOVE HEAT FROM INSIDE THE REFRIGERATOR
TO THE HOT METAL PIPES BEHIND THE REFRIGERATOR.

OVEN

THE OVEN IS THE MAIN APPLIANCE FOR COOKING FOOD, AND IT EVEN BAKES COOKIES!
THE OVEN USES A POWERFUL HEATER TO MAKE THE INSIDE OF THE OVEN VERY HOT. THE INSIDE OF AN OVEN CAN REACH 500 DEGREES FAHRENHEIT.
THE COOK TOP HEATERS ARE USED FOR COOKING WITH POTS - LIKE COOKING SPAGHETTI.

The Little Engineer Coloring Book: How to Build a House

DISHWASHER

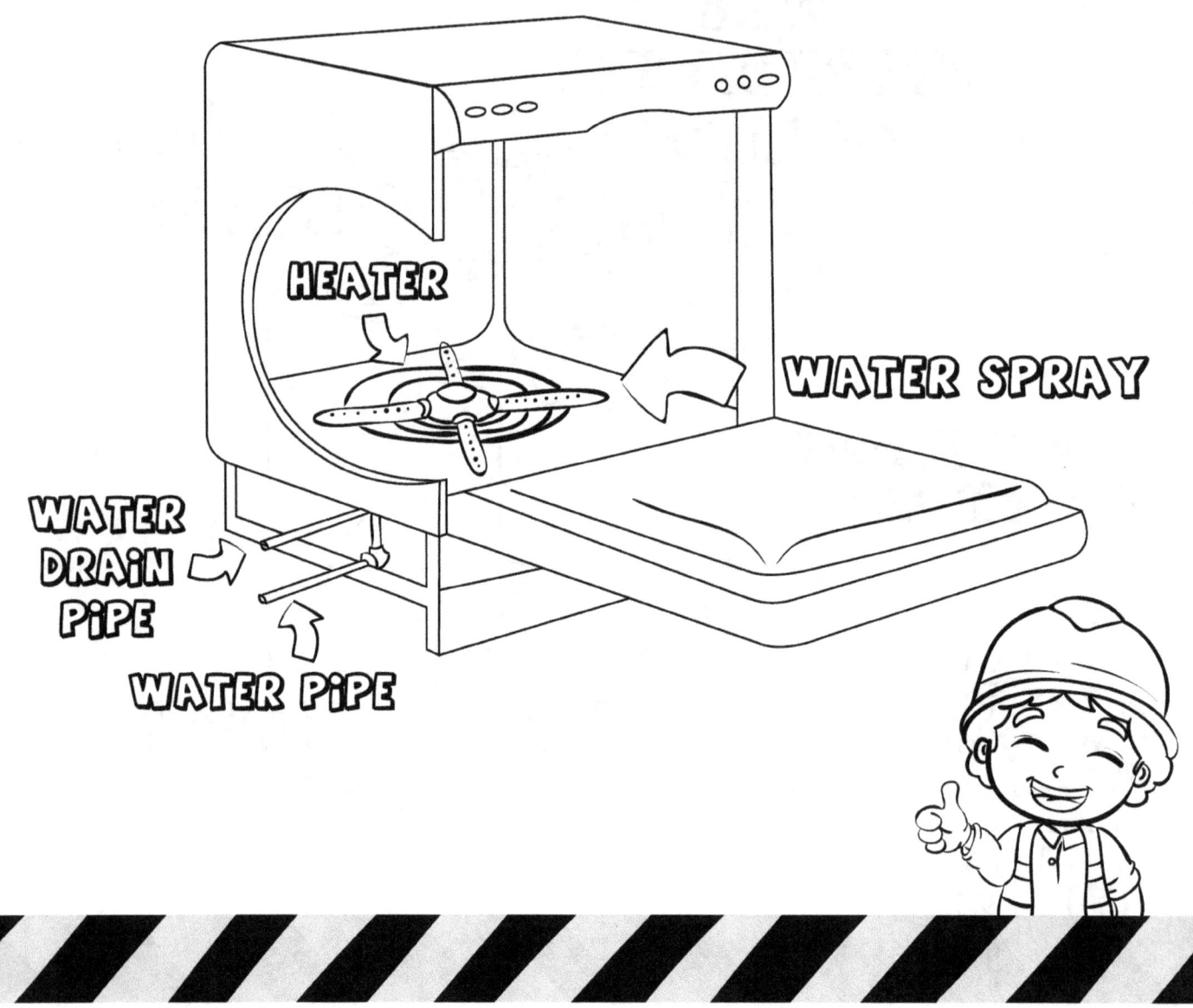

THE DISHWASHER QUICKLY CLEANS ALL THE DISHES.
THE DISHWASHER USES FAST MOVING WATER
AND SOAP TO CLEAN THE DISHES.
THE HEATER THEN TURNS ON AND DRIES
ALL THE WET DISHES.

BLENDER

A BLENDER MIXES FOOD TOGETHER; IT CAN DO THINGS LIKE MAKE A FRUIT SMOOTHIE OR MIX TOGETHER SALSA. A BLENDER CHOPS UP FOOD WITH A BLADE THAT SPINS REALLY FAST! THE BLADE IS CONNECTED TO A MOTOR, AND THE MOTOR IS WHAT MAKES THE BLADE SPIN. A CIRCLE KNOB CALLED A SPEED SELECTOR TELLS THE MOTOR HOW FAST TO SPIN.

WASHING MACHINE

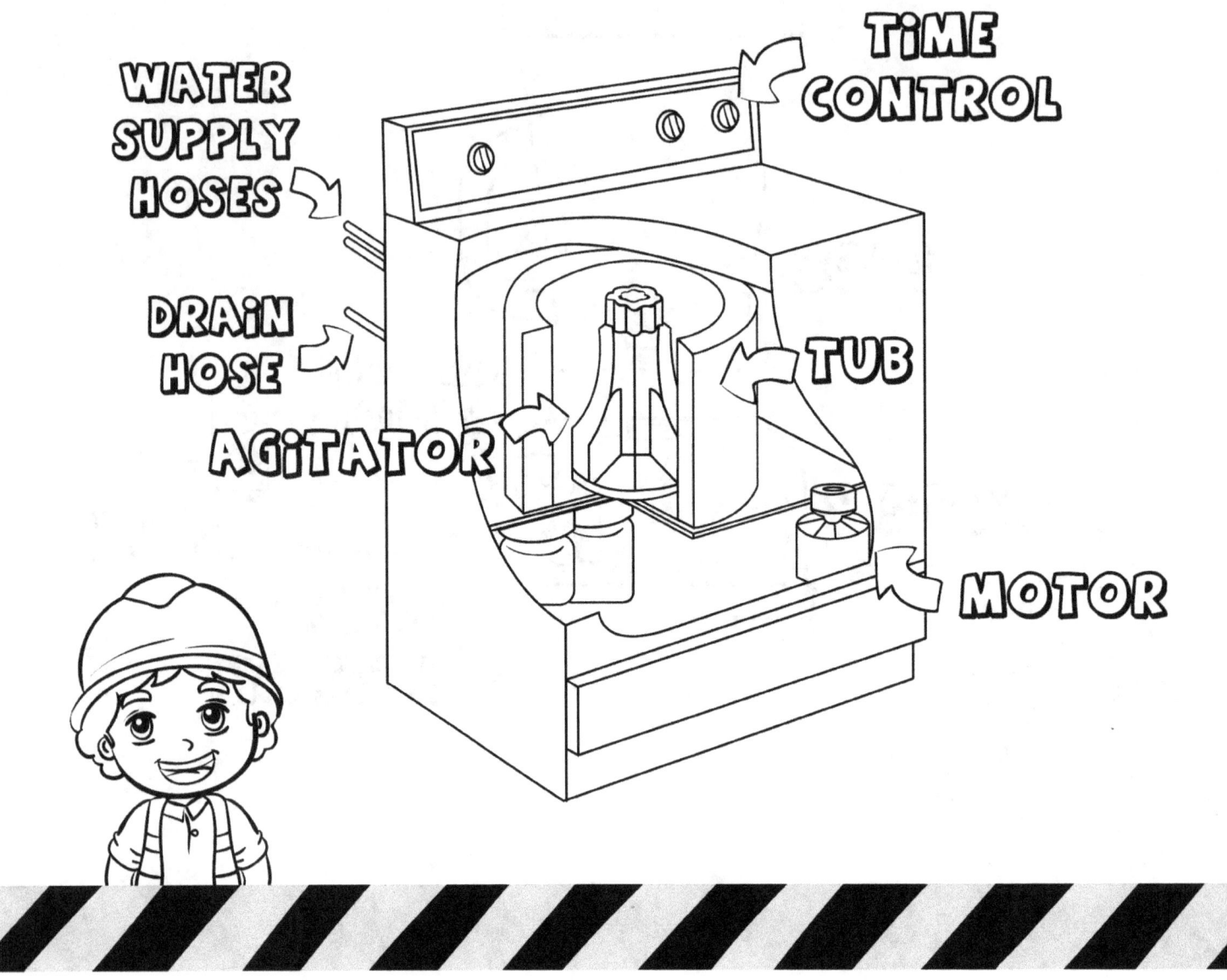

THE WASHING MACHINE CLEANS ALL THE DIRTY CLOTHES.
THE WASHING MACHINE USES WATER AND SOAP
AND THEN SPINS AROUND TO HELP GET
ALL OF THE DIRT OFF OF THE CLOTHES.
WHEN FINISHED, THE CLOTHES ARE STILL WET
SO THEY HAVE TO BE PUT IN A DRYER.

DRYER

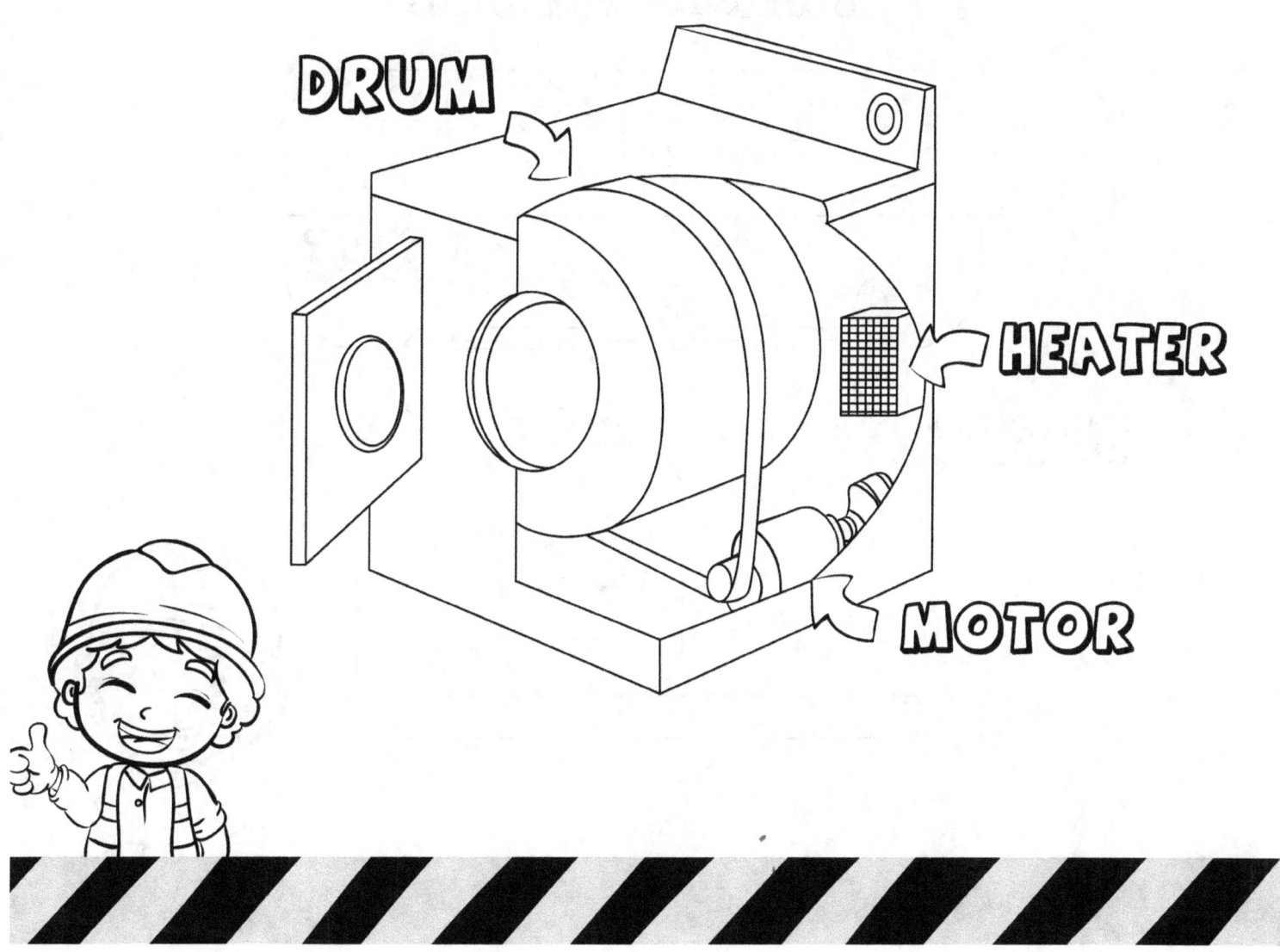

THE DRYER HEATS UP CLOTHES UNTIL
THEY ARE DRY AND READY TO WEAR.
THE DRUM SPINS TO TUMBLE THE CLOTHES
WHILE THE HEATER MAKES ALL THE CLOTHES HOT.

OUTDOOR GRILL

A GRILL IS USUALLY OUTSIDE, AND IS USED TO COOK FOOD. THE BURNER USES FUEL TO MAKE A TINY FIRE IN THE GRILL. THE FIRE MAKES THE GRILL HOT FOR COOKING.

Congratulations! Training Complete

You are now certified and have successfully completed your House Building Training!

Go to: thelittleengineerbooks.com/tle-house-certificate or scan the QR code, and enter your email to get a ready-to-print certificate

SPECIAL PREVIEW

Check out this short preview of another fun coloring book!

LET'S GET STARTED!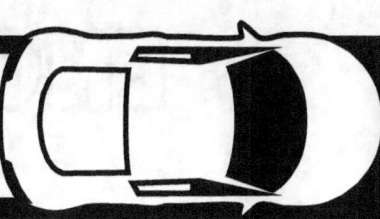

The Little Engineer Coloring Book: Cars and Trucks

ENGINE ACCESSORIES

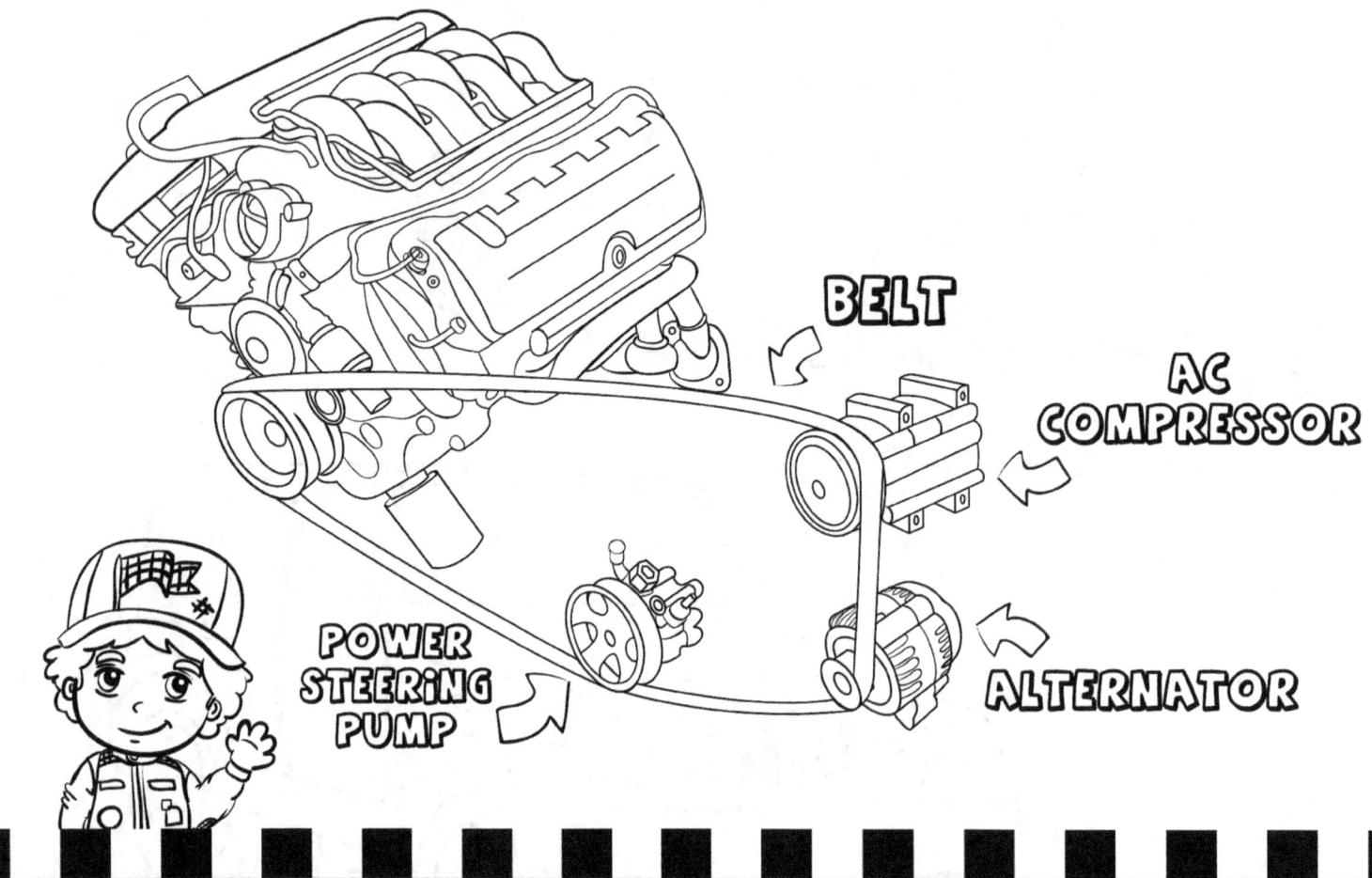

THE ENGINE ACCESSORIES ARE POWERED BY A BELT ON THE ENGINE. THE BELT SPINS THE WHEELS ON THE ACCESSORIES WHEN THE ENGINE IS ON.

- THE ALTERNATOR IS THE BATTERY CHARGER FOR THE CAR. IT IS A SMALL POWER GENERATOR THAT MAKES SURE YOUR CAR HAS PLENTY OF ELECTRICITY FOR LIGHTS, SPARK PLUGS, THE RADIO AND MORE.
- THE AC COMPRESSOR HELPS THE AC SYSTEM WORK SO THE AIR IS NICE AND COLD INSIDE THE CAR.
- THE POWER STEERING PUMP MAKES IT EASIER TO TURN THE STEERING WHEEL.

RADIATOR

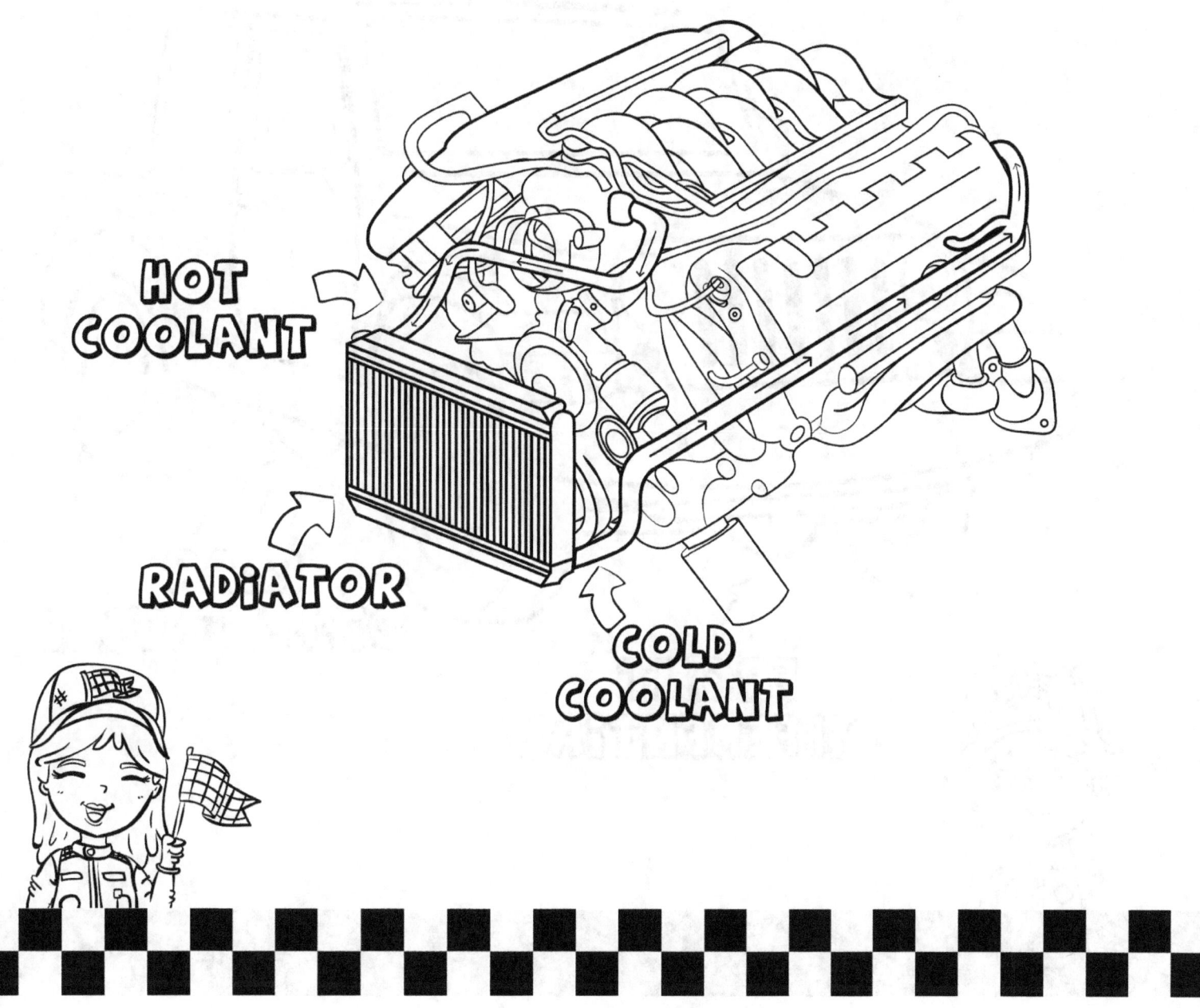

ENGINES GET VERY HOT.

- A RADIATOR HELPS KEEP THE ENGINE FROM GETTING TOO HOT.
- A WATER PUMP MOVES A SPECIAL LIQUID CALLED COOLANT THROUGH THE ENGINE AND THEN THROUGH A RADIATOR.

2 Differentials

FRONT DIFFERENTIAL

REAR DIFFERENTIAL

ON TRUCKS, YOU CAN SOMETIMES SEE 2 DIFFERENTIALS. THIS MEANS THE TRUCK HAS 4-WHEEL DRIVE. SOME CARS HAVE 4-WHEEL DRIVE, BUT IT IS HARD TO SEE THE DIFFERENTIAL BECAUSE THE CAR IS CLOSE TO THE GROUND.

TWIN TURBOCHARGERS

THIS CAR HAS 2 TURBOCHARGERS!

MOST TURBOCHARGERS ARE UNDER THE HOOD AND HARD TO SEE, BUT SOME CARS HAVE THEM STICKING OUT OF THE HOOD WHICH IS REALLY COOL!

FRONT, MID AND REAR ENGINE

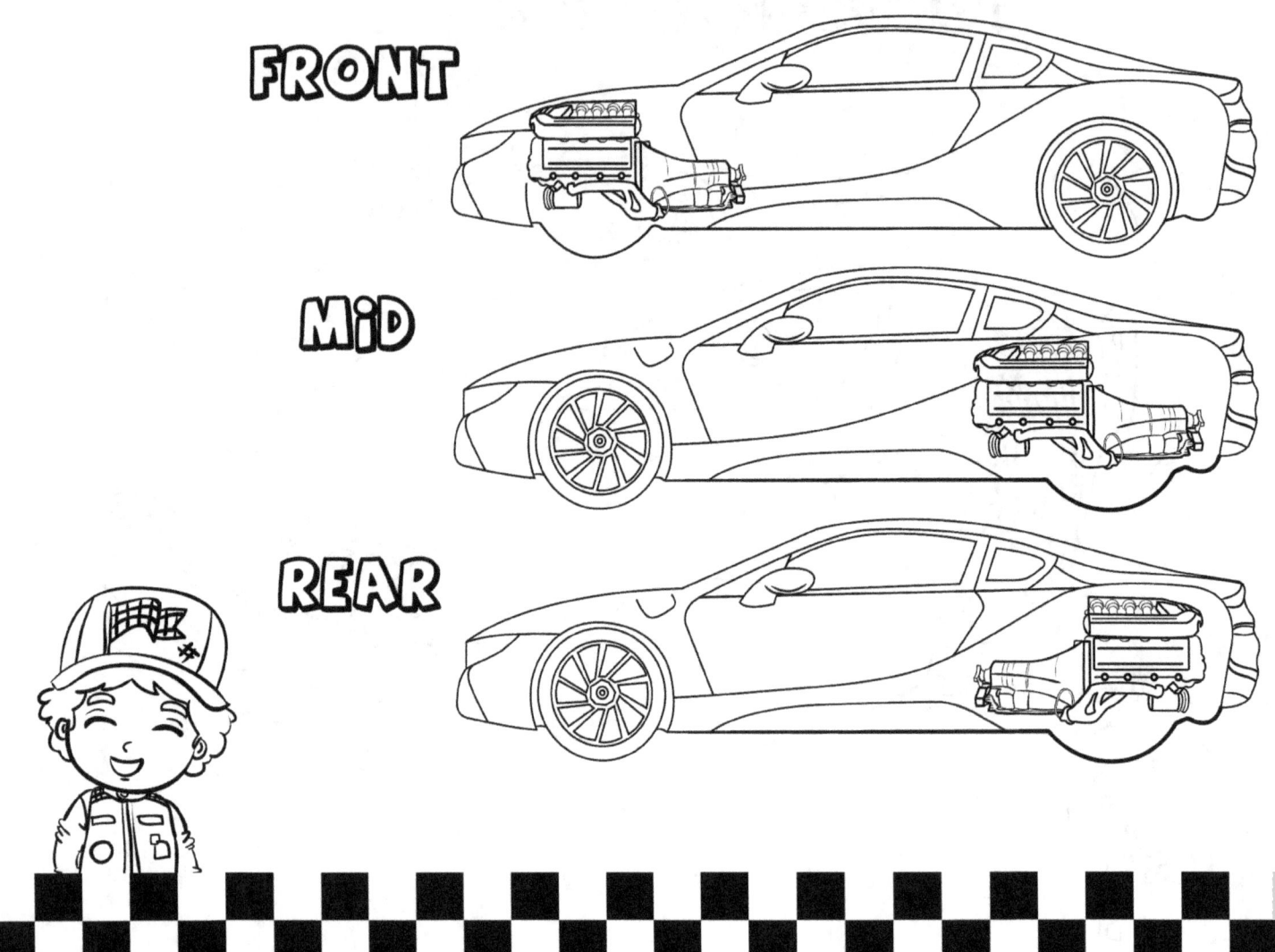

THERE ARE FRONT, MID AND REAR ENGINE CARS.

- FRONT ENGINE CARS HAVE THE ENGINE BETWEEN THE FRONT WHEELS AND THE FRONT OF THE CAR.
- MID ENGINE CARS HAVE THE ENGINE BETWEEN THE FRONT AND REAR WHEELS.
- REAR ENGINE CARS HAVE THE ENGINE BETWEEN THE BACK WHEELS AND BACK OF THE CAR.

We hope you enjoyed the book!
Contact us anytime at CreativeIdeasPublishing.com

We are a US based publisher that consist of parents and teachers. We try our best to make products that our kids will love and we hope your kids love them too!

Ask your bookstore for more great titles from Creative Ideas Publishing!

www.ingramcontent.com/pod-product-compliance
Lightning Source LLC
Chambersburg PA
CBHW081758100526
44592CB00015B/2479